LET'S CODE!

# DEBUGGING CODE

BY SAMANTHA S. BELL

CONTENT CONSULTANT
DAVID A. LASH
ASSOCIATE PROFESSOR OF COMPUTER SCIENCE
AURORA UNIVERSITY

Kids Core
An Imprint of Abdo Publishing
abdobooks.com

abdobooks.com



Printed in the United States of America, North Mankato, Minnesota.
102023
012024

THIS BOOK CONTAINS RECYCLED MATERIALS

Cover Photo: Shutterstock Images
Interior Photos: Shutterstock Images, 4–5, 16, 22, 25 (wrench), 25 (bugs), 25 (clipboards), 26, 29 (top), 29 (bottom); Arief Syauqi/Shutterstock Images, 6; Naval Surface Warfare Center/US Naval History and Heritage Command, 9; Monkey Business Images/Shutterstock Images, 10, 28 (bottom); LightField Studios/Shutterstock Images, 12–13; Chay Tee/Shutterstock Images, 15; Ground Picture/Shutterstock Images, 18, 28 (top); Evgeny Atamanenko/Shutterstock Images, 20–21

Editor: Katharine Hale
Series Designer: Katharine Hale

Library of Congress Control Number: 2023939620

Publisher's Cataloging-in-Publication Data

Names: Bell, Samantha S., author.
Title: Debugging code / by Samantha S. Bell
Description: Minneapolis, Minnesota: Abdo Publishing, 2024 | Series: Let's code! | Includes online resources and index.
Identifiers: ISBN 9781098292799 (lib. bdg.) | ISBN 9798384910732 (ebook)
Subjects: LCSH: Coding theory--Juvenile literature. | Computer programming--Juvenile literature. | Computers and children--Juvenile literature. | Debugging in computer science--Juvenile literature. | Troubleshooting in computer science--Juvenile literature.
Classification: DDC 005.1--dc23

# CONTENTS

People can be competitive when playing video games.

CHAPTER 1

# ROBBED OF VICTORY

Ryan and Emily were playing a video game together. Ryan's race car was far ahead of Emily's. "Finally! I'm going to win this time!" Ryan said, rounding the final curve in the track.

"It's not over yet," Emily replied. She could still claim victory!

A single video game contains thousands or even millions of lines of code.

"There's the finish line!" Ryan yelled. But just as his car was closing in, the screen went dark.

"Oh no!" Ryan said. "The game crashed!"

Emily frowned. "This happens way too often. Now we'll never know who won."

Ryan put down his controller. "I'm done with this game," he said. "Let's play something else."

Ryan was angry that his video game crashed. Like Ryan, many people feel frustrated when their video games or computer programs suddenly shut down. These crashes are often caused by bugs in the program.

## What Is a Bug?

A bug is a coding error in a computer program. It makes the program do something wrong.

Bugs keep a program from doing what it is supposed to do. They often leave users feeling frustrated or annoyed. But nearly every computer program has bugs at some point. **Programmers** try to remove them all. But bugs can be hard to find.

## An Actual Computer Bug

The term *bug* has been used to describe system errors since the 1800s. In 1947, a team of computer scientists realized a computer had stopped working. When they searched for the cause of the problem, they found a squashed moth inside the computer. They taped the moth to a piece of paper. They wrote a note calling it the first actual computer bug.

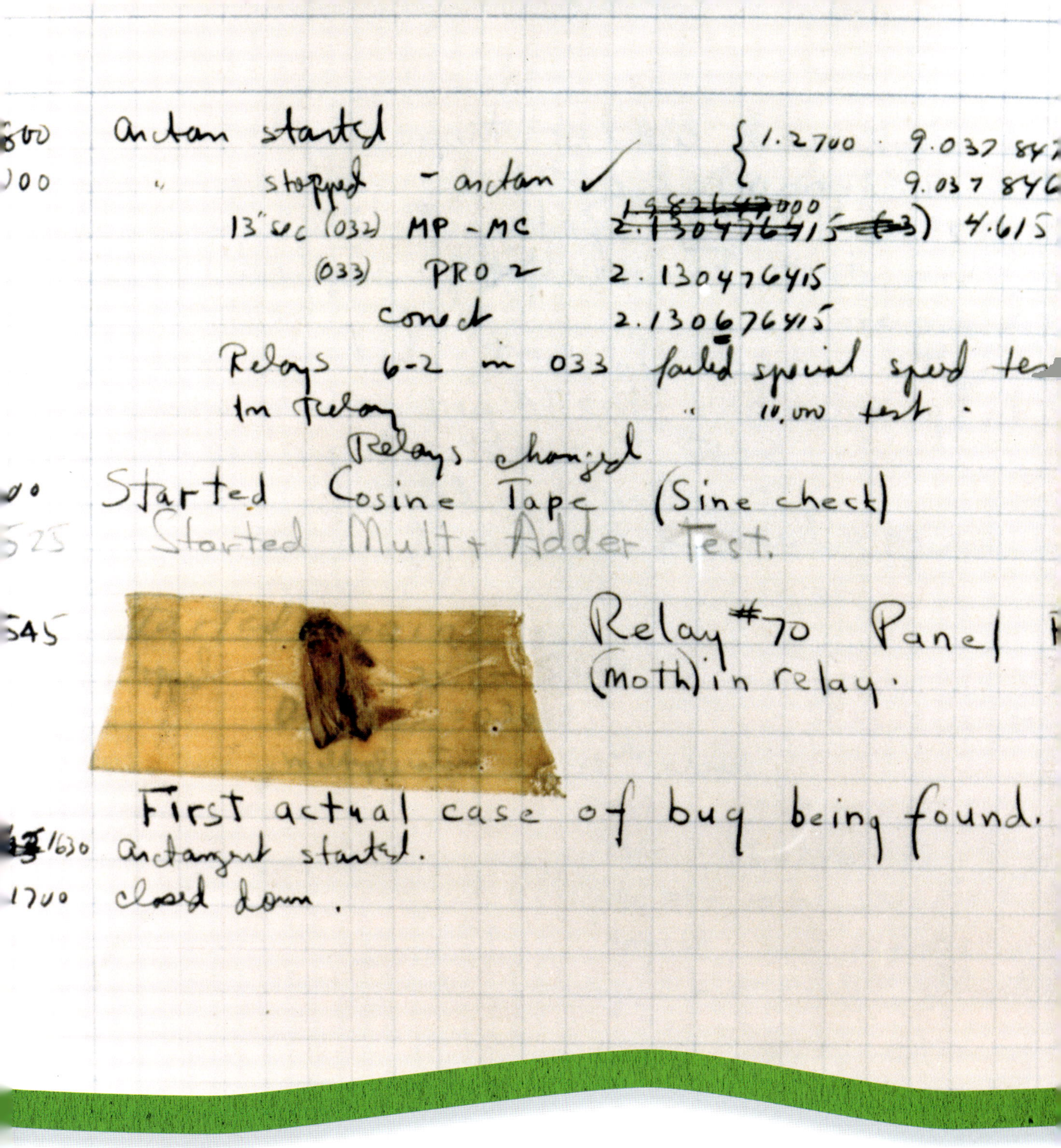


Grace Hopper was a computer scientist whose work led to many of today's programming languages. Her team found a moth in their computer and called it the first actual computer bug.

Programmers have to test their code in order to find bugs and debug the code.

The process of fixing coding errors is called debugging. Programmers try to find bugs by testing their programs. Bugs may also be discovered by people who use a program. Once a bug is reported, programmers can work to fix it.

## Explore Online

Visit the website below. Does it give any more information about coding bugs that wasn't in Chapter One?

### Is a Computer Bug an Insect?

abdocorelibrary.com/debugging-code

Like finding bugs in nature, finding bugs in code can be difficult.

CHAPTER 2

# BUG HUNT

After writing code, programmers test it. Testing helps them find mistakes. During testing, programmers run their programs. They try out all the possible actions users might take. If any part of the program doesn't work, they know there is a bug.

Some bugs are just annoying. But others can completely stop a program from working.

There are many other kinds of bugs. Programmers have to pay attention to find the bugs in their code. Once a programmer finds a bug, they must use problem-solving skills to fix the error.

# PRIMARY SOURCE

Esko Hannula works for a company that helps clients test and publish their code. Hannula explains why a bug-free test is not always good:

> If testers don't find many bugs it may mean . . . that the **software** was poorly tested: the bugs are there, we just don't know it. It may also mean that the developers have done a great job testing their code.

Source: Adrian Bridgewater. "A (Software) Bug's Life." *Forbes*, 22 Nov. 2022, forbes.com. Accessed 3 Mar. 2023.

## Comparing Texts

Does the quote support the information in this chapter? Or does it give a different perspective? Explain your answer in a few sentences.

Successfully debugging a program can feel rewarding.

CHAPTER 3

# DEBUGGING

After finding a bug, programmers try to fix it. Debugging a program can take a lot of time. It may even take longer to debug a program than to write it! This is because it can be hard to find what caused the bug. And fixing an error in the code may create new errors.

People learning how to code must also learn how to solve problems in their code.

There are things programmers can do to cut down the number of bugs in a program. For example, they can plan out the program before starting to code. First, they think carefully about what the program should do. Then they figure out the best way to design it. They think about possible problems. This careful planning helps programmers avoid mistakes. The code will probably have fewer bugs to fix.

## Rubber Duck Debugging

Explaining a problem out loud to someone can make it easier to solve. But it's not always easy to find a person with time to listen. Some programmers keep rubber ducks on their desks. When they have coding problems, they explain the problems to the duck. They tell the duck what each line of the code should do. By explaining the problem, programmers can often figure out the bugs.

## A Good Defense

Another way programmers can reduce the number of bugs is by using defensive programming. Defensive programming tries to avoid possible bugs. To practice defensive programming, programmers **predict** what could go wrong. They plan for those mistakes ahead of time. Then they try to make sure the code will work correctly anyway.

For example, a programmer might create an online form. A user could put the wrong information in the form. This could cause the program to crash. Instead, the programmer plans for this mistake. The program simply tells the user about the error instead of crashing.

# Testing and Debugging

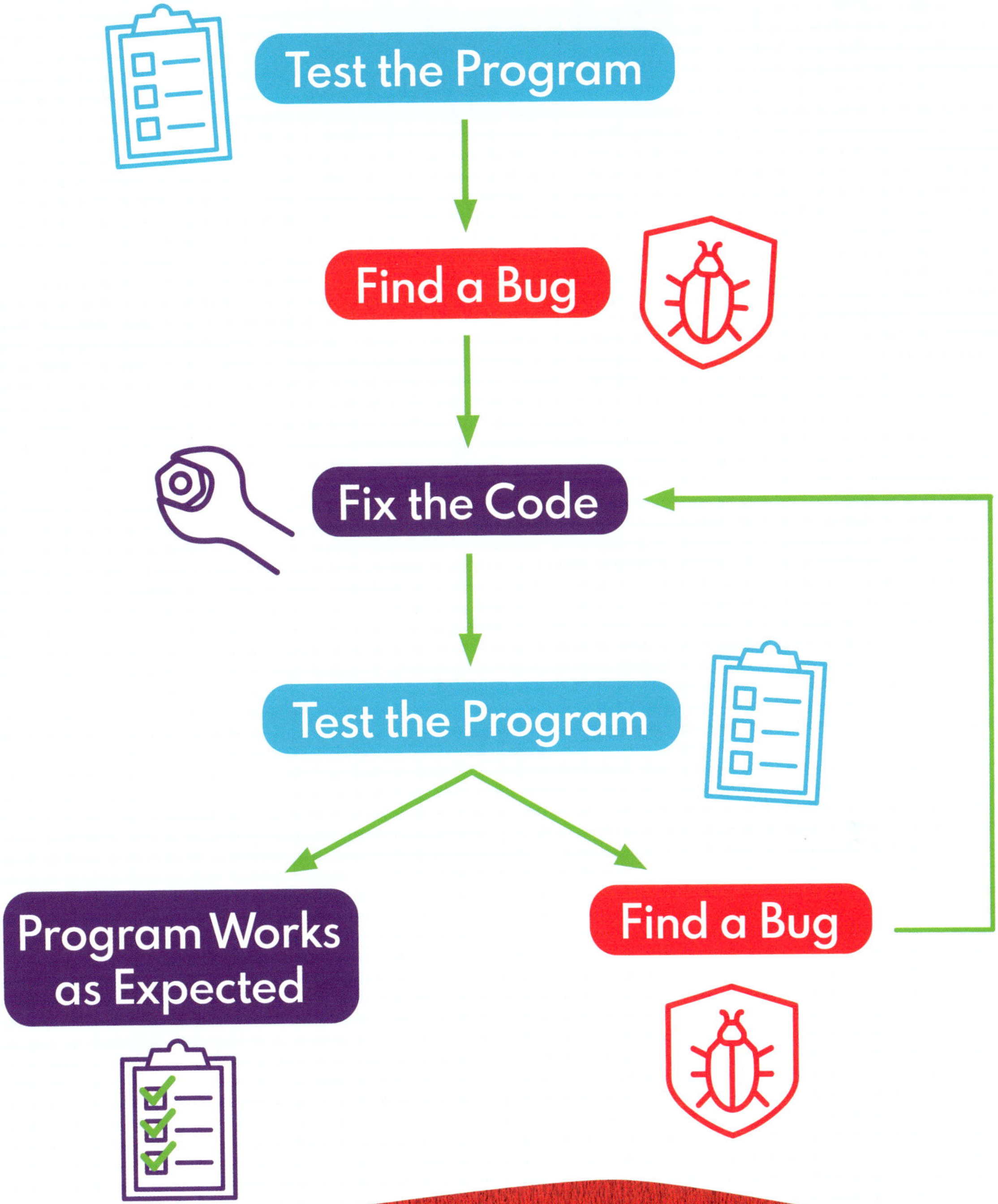


This infographic shows the process programmers use to debug their programs.

Planning ahead before writing code can help prevent bugs.

A bug is avoided. Programmers cannot think of every possible problem. But they can design their programs to handle the most common ones.

Careful planning and defensive programming help prevent bugs. But even the best programmers write code that contains errors. Programmers need to be ready for the challenge of debugging their code. They must become problem-solvers as they hunt for bugs and figure out how to fix them.

## Further Evidence

Look at the website below. Does it give any further evidence to support Chapter Three?

### What Is Debugging?

abdocorelibrary.com/debugging-code

# CODING DATA

Every programmer can make mistakes when writing code. These mistakes are called bugs.

Fixing errors in computer code is called debugging.

Programmers may spend more time fixing code than they do writing it.

Programmers can avoid some errors in code with clean designs and defensive programming.

# Glossary

**blackout**
the loss of all electricity in a city or region

**grammar**
rules people follow to speak or write more clearly, including using proper word order and punctuation

**input**
information that is put into a computer

**output**
information processed by a computer and then sent out, such as text on a screen or sound from a speaker

**predict**
to guess something that will happen in the future

**programmer**
a person who uses programming languages to write code

**software**
all of the programs that give a computer instructions

# Online Resources

To learn more about debugging code, visit our free resource websites below.

Core Library
CONNECTION
FREE! COMMON CORE MULTIMEDIA RESOURCES

Visit **abdocorelibrary.com** or scan this QR code for free Common Core resources for teachers and students, including vetted activities, multimedia, and booklinks, for deeper subject comprehension.

Booklinks
NONFICTION NETWORK
FREE! ONLINE NONFICTION RESOURCES

Visit **abdobooklinks.com** or scan this QR code for free additional online weblinks for further learning. These links are routinely monitored and updated to provide the most current information available.

# Learn More

Bell, Samantha S. *The Basics of Coding*. Abdo, 2024.

González, Echo Elise. *Debugging*. World Book, 2021.

Wood, John. *Code Monkeys Fix Bugs*. Crabtree Publishing, 2020.

# Index



# About the Author

Samantha Bell lives in the foothills of the Blue Ridge Mountains with her family and lots of cats. She has written more than 150 nonfiction books for students. She learned HTML through building websites and wants to learn more programming languages.